THE 5G NETWORK ARCHITECTURE

A Guide That Covers Everything About The 5G Technology

CYRUS JACKSON

TABLE OF CONTENTS

Cyrus Jackson

Copyright

Copyright ©2020, Cyrus Jackson.

Cyrus Jackson

Disclaimer

Though, I ensured the information in this book was correct and accurate as at the time of writing, I cannot guarantee with certitude that it will remain so because of how technology evolves and the dynamic nature of the internet.

Also, the advice and recommendation in this book are based on my experience and this may change due to circumstances beyond my control.

However, I'll try to update this book when such situation occurs and Amazon will send you a **free** copy to keep you apprised.

But the nature of my busy schedules might cause a little delay because I'm really not "**The Flash**".

So exercise patience or simply contact the author at **Cyrus@smartbloggingtips.website**

CHAPTER 1

HOW TO USE THIS BOOK EFFECTIVELY

Hi there, great to meet you!

I want to personally thank you for buying a copy of *The 5G Network Architecture*, a guide that covers virtually everything you need to know about the 5G technology.

I really appreciate the trust and believe you placed on me to give you the needed information about The 5G Network.

I've laid out this book in simple steps with graphics to guide on virtually everything you need to know about the new 5G network technology.

So, if you want to get the best out this book, here's what I recommend you do:

- Take your time to read the whole book, stopping at every chapter to understand each topic and the basic illustration laid out so you don't get confused regarding any topic.

- Re-read the whole book again and you'll discover that you'll gain a fresh memory and the topics that were confusing and vague will be very clear to you.

- Ensure you read the section in **Chapter 17** that covers the meaning of some of the acronyms used in this book so you get a clear understanding of what they do and how they operate.

- Subsequently, use this book as a reference guide whenever you get stuck regarding any topic regarding The 5G Network.

I'm confident that you will learn a lot of valuable information from this awesome book.

CHAPTER 2

INTRODUCTION

5G is the fifth generation of wireless data networks and technology.

The best part?

It'll definitely transform the way network are being used far better than what 3G and 4G combined could ever try to accomplish.

I mean, you'll feel the impact of the 5G technology virtually anywhere you find yourself: in your phone, when you're at home, and in the city you live.

The process of making downloads and streaming online will happen faster than the blink of your eye.

With the 5G network, what seems to be virtually impossible will come true in real life like Cars communicating with each other in other to prevent collisions and accidents.

Cyrus Jackson

You would even be able to download a complete movie under 3 seconds (which is even faster than the blink of an eye).

Yes, that's not a typo. Amazing, right?

This will definitely be the stepping stone for amazing technology advancement and wild imaginations like autonomous driving, internet of things, and lot more.

Can you beat that?

CHAPTER 3

WHAT IS 5G TECHNOLOGY ALL ABOUT?

Now, the million dollar question is what is the 5G network all about?

In other words, what exactly is a 5G network, how does it work, and what's all the hype (what's even special about it) about this technology?

There are 5 brand new technologies that serve as a foundation of the 5G network which makes it possible to carryout amazing functionalities:

- High Bandwidth.
- Millimeter waves.
- Small cells.
- Massive MIMO.
- Beamforming.
- Full duplex.

Let's dive into each of the type of network so you've a good understanding how what they are, how they work, and what they can do:

Cyrus Jackson

High Bandwidth

As good as the 4G network, it can only produce about 200MB (Megabyte) of data per seconds on a very good day.

Now, the 5G network can muster up to 1GB (Gigabyte) of data per seconds which is amazing.

That's about 1000MB per seconds which is 10 times better than then4G technology.

That's very interesting.

Millimeter Waves

It goes without saying that the phone you make use for your daily activities and some of the electronic devices in your homes and offices make use of a specific amount of frequency as distributed by the Radio Frequency Spectrum which is usually less than 6GHz (Gigahertz).

The issue is that these frequencies are starting to get more crowded by equipments using it and frequency carriers can only squeeze so many bits of data on the same amount of radio frequency spectrum.

Which explains why when more and more devices come online; there tend to be a slower network of service and connections dropping rate skyrockets.

Now, how do you handle such a situation like this?

The solution is to open up some new networks which explains why researchers are experimenting with broadcasting on shorter millimeter waves specifically those that fall between 30 and 300 GHz (Gigahertz).

This section of spectrum has never been used before for mobile devices or even electronics and opening it up means more bandwidth for everyone but there is a catch: Millimeter waves can't travel at a faster pace when there are buildings, obstructions, and obstacles hindering their free movement.

In fact, they get easily absorbed by plants and rain.

So, how do we get around this problem?

Cyrus Jackson

Basically, that's where another technology comes into play which is called **Small Cells**.

Small Networks

The wireless network of today rely heavily on large-powered cell towers placed very high above the ground to easily disseminate their signals and frequencies over long distances for usage.

But there's an issue in that kind of deployment because higher-frequency millimeter waves find it difficult to travel through obstacles and building.

What this means is that once you move behind one, you'll most likely lose your signal.

That's where small cell networks solve the problem by using thousands of low-power mini base stations.

These base stations would be much closer together than traditional towers forming a live which is similar to a relay team to transmit signals around obstacles.

This would be very useful in cities because there is virtually no way a user would be able to stick to a particular position.

In other words, as the user moved behind an obstacle or a building, his Smartphone would automatically switch to a new base station in a better range of his device allowing him to keep his connection without shutting him out.

Awesome, right?

Massive MIMO

Simply put, **MIMO stands for Multiple-input Multiple-output.**

Today's 4G base stations have about a dozen ports for antennas that handle all cellular traffic but massive MIMO base stations can support about a hundred ports.

This could increase the capacity of today's networks by twenty two or even more.

Of course, Massive MIMO (Multiple-Input Multiple Output) comes with its own complications as with every technology.

Cyrus Jackson

The cellular antennas used today broadcast information in every direction at once and these causes jamming of signals, dropped connections, and lots of interference which brings us to the next technology called **Beamforming**.

Beamforming

Beamforming is like a traffic signaling system for cellular signals and network.

Instead of broadcasting in every direction, it would allow a base station to send a focus stream of data to a specific user.

This awesome precision prevents interference and is far more efficient.

Which means base stations could handle more incoming and outgoing data streams at once.

Here's how it works:

For example, if you're in a cluster of buildings and you're trying to make a phone call and your signal is ricocheting off the surrounding buildings and criss-crossing with other signals from users in the area as well.

What a **massive MIMO (Maximum Input Maximum Output)** base station does is that it receives all of these signals and keeps track of the timing and the direction of their arrival respectively.

After decoding their location, it'll use signal processing algorithms to triangulate exactly where each signal is coming from and plots the best transmission route back through the air to each phone.

Sometimes, it'll even bounce individual packets of data in different directions off the buildings or other objects to keep signals from interfering with each other.

The result is a coherent data stream sent only to you which brings us to the next technology called **Full Duplex**.

Full Duplex

If you've ever used a walkie-talkie, you would know that in other to communicate, you have to

Cyrus Jackson

take turns talking and listening which is kind of a drag and nerve-racking.

The cellular base stations used today have that exact same hold up because a basic antenna can only do one job at a time.

It's either it transmit or receive which is because of a principle called reciprocity.

Reciprocity is the tendency for radio waves to travel both forward and backward along the same frequency.

To understand this, it helps to think of a wave like a train loaded up with data and the frequency it's traveling on is like the train track and if there's a second train trying to go in the opposite direction on the same track, you're going to get some interference.

Up until now, the solution has been to have the trains take turns or to put all the trains on different tracks or frequencies but you can make things a lot more efficient by working around reciprocity.

Researchers have used silicon transistors to create high speed switches that halt the backward role of these waves.

It's kind of like a signaling system that can momentarily reroute two o trains so that they can get past each other which means there's a lot more getting done on each track and that's a whole lot faster.

The bottom line is that you will need new phones, tablet, portable hot spots …with a 5G radio inside to connect to the new 5G networks. Your current 4G or even 3G gear can't tap into it.

5G networks will also position more data and computing resources closer to you that will avoid blunting 5 G's low latency benefits.

Most of us think of phones when we think of wireless networks but phones may actually be the least interesting thing about 5G because it will also power autonomous cars, so they have awareness of every other car, bike, pedestrian, and traffic signal around them.

Cyrus Jackson

Smart cities based on 5G can make almost anything that's electric as well as connected and aware.

When a bridge needs repair, why can't it tell someone?

With a mesh of 5G connected sensors, it would be able to do that.

And then there's your home, 5G will offer a new way to get internet there. It's skewing cable or DSL and one day, maybe even obviating your Wi-Fi router altogether as devices may just use 5G natively to connect directly to a wireless ISP.

5G has renewed concerns about the safety of cellular radio waves.

Some cities have taken action to block 5G deployment. Some health questions come up around 5 G's use of microwave frequencies and lots of them as we've seen thanks for that large number of small cells that will be installed around us.

While 5G may sound like a lot of microwave ovens mounted on poles running with their doors open.

In fact, microwaves are nothing new because it's, already emitted by your current Smartphone, old cordless phones, your wireless headphones, earbuds and just about anything with Wi-Fi as well as microwave ovens but even 5 G's highest frequencies are considered by scientists to be non ionizing radiation.

You have to move on up to x-rays, gamma rays, and cosmic radiation to find the kind of emissions that will harm cells and 5G is far below that.

5G follows the inverse square law losing power rapidly at even a small distance from the small cell.

The United State Center for Disease and Control says there is no scientific evidence to provide a definite answer to the question of cellular radio wave safety whether it's 3G, 4G, or 5G.

5G may not take too long before it becomes true reality and some investment on your part

Cyrus Jackson

but the complexity of that deployment and debugging should be worth the wait which would deliver the world we've been promised for a long time: where everything is connected, aware, and responsive and a world in which we can stop worrying about the availability of performance of connectivity in the first place

CHAPTER 4

THE 5G CORE ARCHITECTURE

5GS Service-Based Architecture

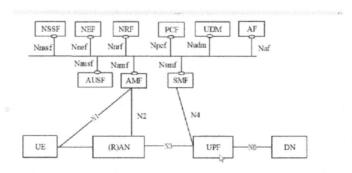

This is the 5G core architecture overview. What exactly is the 5G?

The 5G core is not a technology that exists today or outside of white papers.

While we're a bit ahead of the game here since it hasn't been launched, there is more than enough knowledge of the 5G system protocols for us to have something to talk about.

However, it's prone to change given its infancy state but if you're looking to understand what 5G in the wireless industry is going to mean for the core network, that's what really what I'll be showing you in this book.

I'm going to start with highlights of the EPCs deficiencies.

EPC means **Evolved Packet Core** and the previous architecture and the reason for that is because with any new technology, the question is always what is the innovation and I think it's going to help by going over stuff about the **Evolved Packet Core** that didn't age particularly well or that we don't particularly like or at least that we can see that there could be a way to do it better and you're going to see that in some of the examples I'm going to bring up that the 5G core is really going to target and improve upon these.

There are going to be a fair amount of new terminologies and new acronyms in particular in this book which is unavoidable when a new architecture comes to existence.

It goes without saying that it's going to bring its own terminologies which you would clearly see with the 5G core which is not too bad.

I'm going to go through the overview of the key components of the 5G core system which is 5GC and 5 GS and that is the same distinction that the wireless industry drew in the Evolved Packet Core.

In other words, in the previous architecture, we had the **EPC (Evolved Packet Core)** which was the core network specifically and the EPS which was the end-to-end service architecture.

The same is true in the terminology of the 5G core which implies that 5GC is core while 5GS is end to end and service architecture and I'm going to dive into the 5G service based architecture at the SBA which is kind of a different way of thinking about how systems talk to each other.

I'm going to talk about that new architecture: the SBA.

To start with, the **Evolved Packet Core** problems. This is the Evolved Packet system here:

Cyrus Jackson

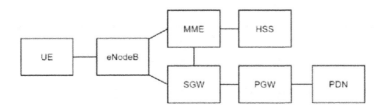

Which is of the bare-bones drawing of the EP of the EPS and the core is really these four components **(MME, HSS, SGW, PGW)**, if you are not familiar with this diagram, I would encourage you to carefully study the diagrams because I'm going to be constantly drawing comparisons between this and the new architecture and read up on the EPC.

While I've written on other topics that specifically covers EPC and goes over what the systems are and how they operate, it would be more beneficial to you if you first master or at least understand the basics of Evolved Packet core in this book.

CHAPTER 5

EVOLVED PACKET CORE PROBLEMS

The pattern I implemented in writing this book is to present the 5G core and draw comparison with the packet core and its four systems and the first problem I wanted to focus on here is:

1. Too Many Protocols

EPC Problems

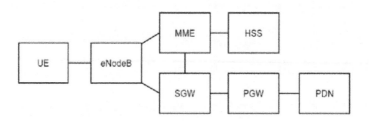

So the image above is just an example of an MME that you could draw similar for the other functions but it's just rather than having one signaling protocol that does signaling well once, what we've kind of ended up with the

Evolved Packet System is that for every new system that gets introduced, there are always every new interface that's needed between two new systems, it always seems to introduce its own signaling protocol.

So **SLSAP** exists only between an **MME** and an **SMLC**. The **SGSAP** is a signaling protocol that exists only between an **MME** and a **VLR**.

What exactly is a signaling protocol?

Simply put, it's just an application that conveys attributes and values between two systems.

So, if you have a protocol that can do that and does it well with bells and whistles features (like the ability to route at an application level and the ability to maintain state machines and sessions as needed between systems), that's really all that assuming protocol needs to do so.

If you've it done once and you've it done very well, you can leverage that across all the 3GPP interfaces that we have for signaling and get rid of this unfortunate trend that we've ended up with where every new application ends up being its own signaling protocol.

That helps us in a bunch of ways. If we reduce the count of signaling protocols that reduces what a new software developer would need to develop if they're going to build an MME from scratch to compete in that market.

That actually reduces operational complexity for operators that need to support an MME because it means you don't need to be proficient in a great number of signaling protocols.

In other words, if you know the one signaling protocol they're using for all signaling, it's operationally much easier and simpler.

So, this helps everybody since we have fewer sealing protocols as opposed to more and that is definitely what the 5G core is going move towards.

2. Closed APIs

Cyrus Jackson

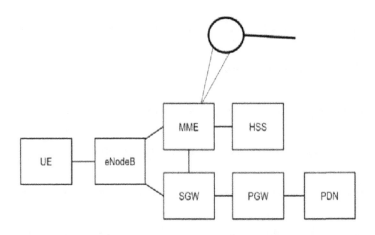

Even though, the way these systems communicate to each other is standardized, the way that an **MME** will query for subscriber data.

For example, HSS is standardized but the way that an operator would ask something like: "Tell me about the sessions that this subscriber has this MME or tell me about all the PDN connections on this P gateway and how you achieve that right now when the Evolved Packet System is all based on the particular vendor of the particular system you're talking about.

Different vendors are going to have different user interfaces and different APIs supported in their platforms to get that kind of data but there's a movement towards open API.

With an open API, that standardized by 3GPP, you can have it and the 5G core does have it that rather than this per vendor approach, you have a single way to get data to request changes and for data to subscribe or be notified of changes to data.

All of that's going to be standardized with an exposed open API that's going to make these a lot easier to manage and it scales better, it supports multi vendor environments much more easily but for closed API, you'll need a certain skill set and proficiency in every single vendor platform that you have and then from an integration perspective, if you get say an analytic system that's going to grab data from different vendors and all of their subscriber data, then all you need (all of a sudden) for that analytic function to support all these different vendors, it becomes much more complicated but with an open API, a standard way to get

Cyrus Jackson

standardized types of information from all of these systems becomes much easier all around and that if that is something that's being moved toward for 5G core.

But Evolved Packet Core is all closed APIs which I definitely don't like about it.

3. The Gateway Selection Isn't Flexible

This drawing and every color in the image below illustrates a network topology that you might potentially want but for all these X's, their topologies that you can't have.

EPC Problem 3: Gateway Selection is Inflexible

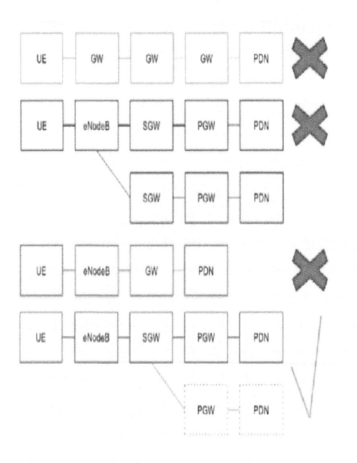

Architecturally, it is required that you always have GDP speaking gateways between a user and a destination network. That's the UE and PDN (Packet Data Network).

Cyrus Jackson

So in a scenario where you've got three gateways: the orange colour in the image (that's not possible if you ever needed or wanted to extend a third Gateway.

Architecturally, Evolve Packet Core does not have a possibility of three gateways which means it's an S gateway talking to a P gateway always.

The second line in the image above is invalid because you've got two S gateways for the same UE so you can have two S gateways but always for different users, you can't have one S as well as you can't have two S gateways for two different IP sessions for a single subscriber at a single point in time.

You may for whatever reason want to do that at some point but architecturally that's not possible.

The third line is probably the most practical. If you just want one gateway, you've got a small deployment and you want one subscriber to one gateway to reach the internet.

Again, this is not physical but logical architecture which means you must logically you've an S gateway and a P gateway.

There's this forced dichotomy of the gateway which serves the access that is the **S gateway** and the Gateway which serves the destination that is the **P gateway** and they must always both exist for any data session in the Evolved Packet system.

So the green at the bottom is the only valid scenario even if you only wanted one gateway, you would logically have to have two in terms of the architecture in other for the surface to work.

Again, with the use of the logical architecture, the pink is invalid and my point in this picture is that the Evolved Packet Core pigeon-holes you a bit in terms of how you can manage gateways and how you can manage different roaming scenarios.

The 5G core is going to kind of rethink this. Specifically, if they're going to get rid of the concept of a serving gateway in a packet data network gateway and just have the generic concept of a gateway or a user plain function

Cyrus Jackson

and have no limit on how many users plain functions can be involved.

This could be as few as one and as many as number.

In other words, three gateways would be possible for the most part except for the multiple ask gateways in the 5G Core.

4. Session Management Is Complex

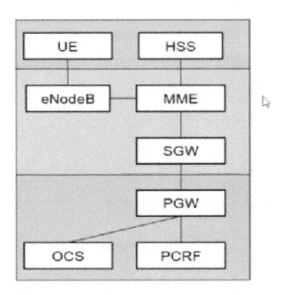

Session management which is a very difficult thing to illustrate in a book but if you've ever managed or if you've ever had to kind of design for the Evolved Packet system with multiple SKUs of sessions, it does become a bit difficult to manage Session Management and Evolve Packet System.

So what happens is for decisions like what QoS is appropriate for a session from the phone towards the destination IP network or what APM should be selected or whether to invoke a policy control server and if so which policy control server or whether to invoke an online charging server?

And if so which online charging server and under what conditions, right?

So all of these questions need to be answered because they are all fundamentally session management questions: whether to apply policy control for a session or not, whether to apply charging to a session or not and what ends up happening is you get the requests coming from a user equipment.

These equipments can ask for QoS levels and they can ask for an APN by name or not or ask by IP version or not.

There's as much stuff they can or may not ask for since you're potentially limited by either the energy or the tracking area for an MME perspective in terms of what that radio is capable of since the MME can take both of those into account and can also takes into account from the HSS what the subscriber is provisioned for and then takes his own local configuration into account the MME.

For example, you might have different roaming agreements where an MME will be the enforcer to say well this Sims II range, I only allow up to QoS to CI value seven but for these other ones, I allow up to five or I've got volte agreements with these ones.

So, I'll permit all the way down to QC i1 and I'll set the VOPS flag a certain way.

All these decisions are being made in an MME, then the MME is going to formulate all this into a request to the other operators P8 which is assuming roaming otherwise all the boxes in the image above would all be green out to the P gateway where the MME serves a session management for the access network and the P gateway speaks for the home network.

The P gateway takes all these inputs from the MME and this is limited to what the MME decides.

There's no standard way to decide whether or not any of these things apply and whether or not to invoke a P gateway.

Cyrus Jackson

There's nothing standardized that says MME requires a P gateway to do something like that.

So you end up with a lot of ad hoc per hop decisions being made by individual systems at the end of the day that desires to get a predictable result for the subscriber.

Session management is done in the Evolved Packet core. It's not that it can't be but just that it hasn't grown well which has made it become a bit complicated and how it gets deployed.

As networks become more complicated over time and as you have more and more SKUs of data sessions to manage.

If you have multiple charging systems and multiple policy systems towards the internet with common APNs and then it becomes okay.

The next question that comes to mind is: How do I decide which policies to apply?

How do I get the right information to a policy server to make decision X about this type of subscriber or decision Y about these other type of subscriber?

They will seek to manage session in a better way and some aspects of mobility management as well in a better way and evolved in the 5G Core.

5. DNS Becomes Complex

Just because you're adding a lot of service information and a lot of logic for an MME to make ultimate decisions about how to select S gateways and P gateways will reflect on the 5G core because what's going to happen in the 5G core is they actually expand the scope of DNS in the 5G core to include not just gateway selections but discovery for entire Network topologies for every type of network function and priorities and weighting and stuff for any type of network function and not just gateways.

So, with that expanded scope, my impression anyways, they gave up on the idea of just expanding and expanding DNS to deal with everything and they defined their own data structure into what's called the network repository functions which I'm going to dive into in another chapter.

Cyrus Jackson

This just becomes a bit too complicated to manage especially as they grow the scope of it which explains why they have ended it in the 5G core.

CHAPTER 6

One of the big picture of the 5G core is that it gets rid of most wireless sealing protocols and that is big change. This means that diameter is gone. **GTP** is gone. **DNS** is gone. **GDP** control is gone.

Which means that it gets rid of all of those outdated features in favor of **HTTP 2** which is a newer version of **HTTP** (which most people are more familiar with) and more compatible with the latest technology.

HTTP2 came in 2015 which is a fairly new protocol version for **HTTP** and with **JSON (JavaScript Object Notation)** which is the application layer in HTTP messaging.

So, one signaling protocol stack for several places in the **Evolve Packet System** which provides an open **API** as I alluded to into all network functions allowing redirect requests toward any of them.

Cyrus Jackson

It gets rid of the concept of gateway selection in general with that in favor of network repository which I mentioned and it removes session management from the **MME and P gateway**.

So, what it's going to do is that it's going to take the **MME** and break it up into several parts.

I'm going to give a better illustration in the next chapter for you to understand what I'm talking about but one of them is the session management function where instead of an MME talking to a PIA gateway for session management.

In other words, you've got the session management function in that dedicated role in the two operators (if you're talking about roaming or just want if you're talking about home case) where they both speak for their operator for session management in general.

The big picture of the 5G includes:

- It gets rid of most wireless signaling protocols and replaces them with HTTP2 transport and JSON (Javascript Object Notation)

- It provides an open API into all network functions allowing read/write requests to any of them.

- It gets rid of the concept of gateway selection and replaces it with the concept of a "Network Repository", with a much broader scope.

- It removes session management from the **MME** and **PGW** and places it centrally in the "SMF function, where sessions can be negotiated between SMF's for home and roaming cases.

Cyrus Jackson

CHAPTER 7

REFERENCE ARCHITECTURE OF THE 5G

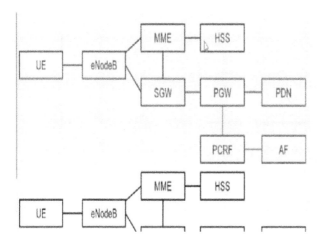

The image above represents the 3GPP reference points in the 5G core.

Although, this is not all of them but it basically covers all of them in the base architecture that are included in TS23501.

What they've done is they've taken the S gateway in the P gateway and put them as just idea of a packet forwarding gateway called the user plain function and had N number of them.

With this N 9 interface, you could have multiple UPFs as needed where they signal to the control part which is the session management function which is really the control part of a P gateway, the control part of the N gateway, and the control part of an MME for session management all rolled into one.

So, they took a part of all three of those and put them into session management function. DN is Data Network which is equivalent to a PDN (Packet Data Network).

Let's dive into a direct comparison of the two to give a better understanding of how it works.

The User Plane Function is taking the user plane part of the S gateway.

There's a control plane part of the SEP and there's the user plan part.

Cyrus Jackson

What they've done is that they've subtracted that which is the security anchor for a subscriber the a USF from that which is a the mobility management anchor for the subscriber which is now the AMF whereas previously the MME played both roles and exchanged security info between MME over an S10 interface.

So, they've removed authentication apart from the MME and put the MME as just the access and mobility management function.

CHAPTER 8

5GS SERVICE-BASED ARCHITECTURE

5GS Service-Based Architecture

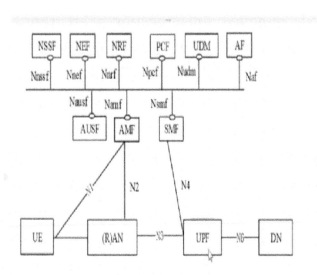

This is the service-based architecture and you're going to notice that this is most of the same names as what I had in the in the picture above for the reference point diagrams where what they have done is that it is the same the same network but drawn in a different light.

Cyrus Jackson

For example, you'll see in the reference point diagram that there's an interface between the AMF in the SMF called N11.

Well, N11 is an HTTP new request to the HGB 2 interface of the SMF and so HTTP 2 interfaces have their own name which is just the letter N followed by the network function name.

So, the SFF interface on the SMF.

So instead of thinking of it in the traditional telecom way of systems with lines between them, think it of it in terms of network.

In other words, think of this whole thing at the top of the diagram above as just an HTTP to network which explains why any system on the line can potentially talk to any system that's close to it including the abstract concept of an application function to talk to these other systems.

What this means is that there's a standard way to request from an AMF for whatever information that you're going to request rather than having it oriented all on what the

The 5G Network Architecture

particular source is or what particular destination is or for what particular type of request.

It's all oriented on an interface and what requests can come into that interface. So, it's a service based architecture and anytime you see by convention in the 5G core or anytime you see the letter in capital followed by a lowercase series of letters, what that means is that series of letters is the name of the network function rather than having an with a number.

You'll also see at the bottom of the SBA that the RAN of that interfaces are kind of not a part of the SBA but are in one, two, and three but they are not HTTP2 interfaces and 3S parties are playing as N6 but the RAN in particular was kind of the odd man out and is not going to support HP as at the time of writing this book where it's not really part of this otherwise core specific service based architecture.

In other words, the 5GS service-based architecture:

- Uses HTTP2 protocol to replace all diameter, DNS, and most GTP interfaces.

- One common HTTP2 interface on a network function (NF) to be able to support any type of request to that NF. By convention "N {function name in lower case}" refers to an interface into the HTTP2 stack of that network function. For example, "Namf" to access AMF services, regardless of the context.

- Provides for common procedures and session state machine logic across all sessions between all types of network functions.

- Provides for TLS encryption optionally. In practice, all web browsers only support H2 (encrypted) and there is little to no support in the IT field for clear-text HTTP2 (h2c).

- Uses application type JSON (Javascript) within HTTP2.

CHAPTER 9

PRODUCTION REDUCTION

Protocol Reduction

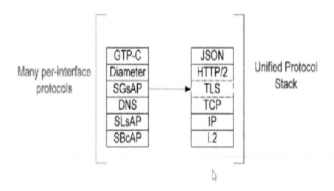

While the image above can be a little misleading but it's basically taking all six of these and you can actually add more signaling protocols that existed in the involve packet core and they are all being replaced with this single stack.

Cyrus Jackson

This is just the 3GPP picture of the same thing which explains why this is the way it's written right now for HTTP2 concepts.

Still, officially in 3GPP dotted line around TLS but again there are open questions in terms of practical vendor deployments regarding if vendors are going to deploy or even recommend deploying with TLS or without.

A one possibility that I see is that you could end up with unencrypted within a network and an encrypted between networks you could also end up with the current web model which is just everything uses TLS.

The 5G Network Architecture

CHAPTER 10

HTTP2 Concepts

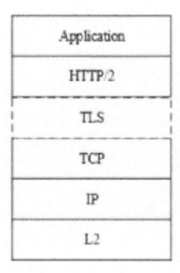

Some highlights about HTTP2 as a protocol is that it supports multi streaming over TCP which means it's not vulnerable to what is called header blocking and lightening blocking in HTTP.

Cyrus Jackson

So, HTTP hit a limit on how many requests could be in wait before getting responses or before new requests can be made.

Well, with multi-streaming HTTP2 doesn't really have that problem since it basically takes a lot of SCTP concepts which is a transport layer protocol and kind of applies them to HTTP.

It supports pre-emptive data pushes which means that you can have one request that an answer where the answer gives you both the answer to was being asked and the answer to what might be asked in the next question which is something they should make which previously was not supported in HTTP.

One of the biggest changes is that we're moving away from niche signaling protocols that are specific to the telecom domain to a protocol that is widely supported in the IT industry today and has a significant development base.

So, web developers around the world are going to support HTTP, HTTPS, JSON and work with them regularly where you have readily available free Apple downloadable app appliance service that you can get that support and all of these protocol stacks.

Within the course of an hour, I was able to take the 3GPP schema network function and a JSON schema.

Copy it and paste it directly into an open JSON server schema and load with an HTTP to a passion install and it'll get to a point where I could actually do working requests to a working network function in the 5G core and I was also able to get it to a functional network that could actually take real requests and give real answers where all that was remaining was just to define the application level intelligence under the hood but to get the actual protocol stacks working, it required very little effort from my part compared to before and that effort is shared with huge potential for synergy with the rest of the IT industry that's where we're moving toward with this move to HTTP2.

Cyrus Jackson

CHAPTER 11

NETWORK REPOSITORY FUNCTION

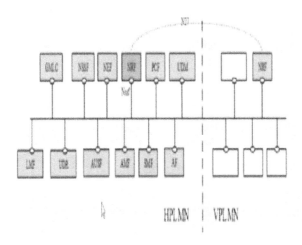

So, the network repository function concept is going to be replacing DNS for Gateway selection essentially for the 5G core.

So, network functions registered themselves with their services and IP information in a data store called the NRF.

Basically, the idea would be rather than you going to a computer and configuring DNS records for MME at the AMF (which you can see at the bottom of the image above) when it connects to the network, it's going to register itself with the network repository and say: "Hi, I support all of these services and interfaces, here's my IP interfaces and my IP addresses that relate to these services that I'm saying that I can provide for the network".

So, other systems will connect to the NRF provide similar results.

For example, if the AMF needs to talk to an SMF, it could actually get and pull data from the network repository and say: "Hey, I need to talk to an SMF. What SMFs exist in this network can learn about their IP information and you're off to the races.

So, rather than doing something specific to Gateway selection in DNS, you have this much broader network discovery concept that's possible between systems inputting their own data into the NRF to be retrieved by other systems that need to talk to them which reduces the number of places where data means to be defined and it also reduces the potential for human error and also increase the scope of what can be learned and what can be done

Cyrus Jackson

between systems in the core by replacing DNS with this repository concept.

So, I'm actually quite excited about it since it's similar to dynamic DNS but it is something that's done very differently and I think a much scalable way than what the DNS service was doing for us for Gateway selection in the Evolved Packet Core.

In other words, it's kind of DNS on steroids which is used for full network topology discovery and it's very similar to DNS in the Evolved Packet Core: the MCC and MNC.

That is the P element of an operator is used at the top level of its data store hierarchy so that lends itself well to roaming scenarios in which your operators (MCC and MNC) is uniquely yours if your NRF can be uniquely discovered, then other operators have a unique reference to get to all of your operator data and then you've got whatever security you would want at different levels of hierarchy to permit what operators to query and who should receive it.

CHAPTER 12

EXPOSURE CONCEPT

The new network function called the NEF or Network Exposure Function exposes or makes available the whole 5G system and its network functions as kind of a point of contact.

So, it supports (at the time of writing this book) three types of operations:

It can take monitor requests to subscribe and be notified of state changes for sessions that exist in other network functions.

It can provide requests to change data in network functions and actually defined our request type to invoke and the NEFT requests policy and charging changes which is potentially useful for analytic tools to subscribe for data types or to pull data.

So, it gets you access to more data from more systems in a more generic way that's just going to lend itself to big data and analytic tools and give more potential applications for them.

Cyrus Jackson

It's kind of more data empowers big data for sure which is potentially useful for troubleshooting as well particularly in roaming scenarios.

The reason I say this is that it gives you a way to get some types of information that you can't really get manually in roaming scenarios today like if you had operator A trying to support a complaint of poor service or something from a customer of operator A while they're roaming and operator B.

Well, there are advantages in knowing what the MME knows about a subscriber or whether it's being rejected by the MME but you can't find the home operator because it's the visited operators MME.

So, that triggers somewhat manual processes between operators but the NEF gives you an ability to query and subscribe for data within a given operators PLMN. In concept, they could be given the rights.

You could have permissions between operators that say: "From my own MCC, MNC, or my own subscribers, allow me permission to query for the AMF session state and mobility state of a subscriber and the SMS session management state of a subscriber".

So, that you can learn as the home operator about your own subscribers even as they roam to other operators through the subscription concept to the visited operator and that's not something that's really possible in the Evolved Packet Core architecture today but more generically this exposure concept allows for non 5G core systems to get information from the 5G core in a way that's not standardized today for anything really equivalent.

Cyrus Jackson

CHAPTER 13

NETWORK SLICING

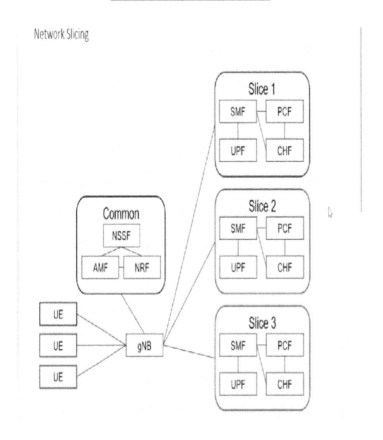

Network Slicing

Think of a network slice as a session management profile.

So, should the book policy control: yes or no?

Should they invoke session management: yes or no?

Should you invoke charging functions: yes or no? To which particular charging function?

All of that can be determined through a session management profile concept called a network slice.

Sometimes, this is illustrated more virtually. So, a virtual PCF that gets invoked in network slice 1 but doesn't get invoked in network slice 2.

Which means you might have one PCF and multiple virtual PCF that treat the requests differently based on the network slice.

To me, that's really just a concept of having a profile for policy control and you don't necessarily need to have multiple physical boxes or even necessarily multiple virtual boxes like a virtual SMF to handle different requests without knowledge of the others.

Cyrus Jackson

It could essentially be just the network slice is a profile ID in that map's to all the logic that we want to provide for subscriber.

With this, you have Network Slice selection function which determines what networks slice is appropriate.

So, an AMF is not broken up into slices neither is the NRF or the NSSF but the session management function can have different logic for different types of subscribers as can the policy control function and the charging function and they can go potentially based on being part of different network slices.

The way the actual numbering is formatted, you'll often hear this long-winded acronym **SNSSAI which stands for Single Network Slice Selection Assistance Identity.**

In layman's terms, it's the profile number for a network slice and that profile number can be provisioned and communicated to the UE and the UE can know the kind of network slices it has access to and it's written in a granular way.

Again, if you relate this to a roaming scenario, you can have it so that the first part of the network slice is a common format so you know the difference between commands like: "Treat it like internet or treat it like IOT or treat it like X, right?

So, the first part defines that the surface profile that you would want at a visited network level and then you have a more granular field for more home operator specific kind of specialty pieces.

In other words, you don't have policy control in the visited network that's always in the home network which means you can base policy control and charging control on a more granular field in
the network slice identify or the SNSSAI whereas the more generic logic to be followed by the visit operator takes the first kind of common standardized formatting of a network slice.

Network Slicing is a bit more prone to change as my impression from the specs as well as a few TR is already asking how we deal with the problem that are emanating from its processes.

Cyrus Jackson

So, it goes without saying that the details of Network Slicing may change a bit over time since it's at a high level and it's just the idea of taking a single Network and logically dividing it into slices of network to provide for different subscribers.

CHAPTER 14

RAN CONSIDERATION

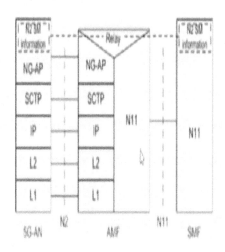

The NIR which is the new radio is not part of the service based architecture.

So, when I say SBA, that's core specific and not part of the 5GS.

Which means that there are no HTTP2 interface in the RAN.

Cyrus Jackson

I'll also note that it's the only part of the whole system that preserves the SCTP transport protocol for OSI layer.

Everything else moved to TCP which is part of the HTTP stack but the access network including access network towards the AMF but it includes SCTP transport protocol and not TCP. It's the one exception to the whole architecture that uses HTTP.

ERAB concept extends to the user plain functions which is equivalent to ZOE. ERAB was an EPS concept to the EPS radio access pair.

Bottom line is that what used to be the S gateway to you as a needed B supporter. S gateway is being replaced with the user plain function or a UPF and if you're used to talking to an MME, the MME is being replaced with an AMF and so the S1 interface is basically replaced with N3 in the case of user plain and N2 which used to be towards an MME is now towards the AMF.

The image above is from 3GPP and it shows the user plane protocol stack which is unchanged.

If you look at it the Access Network GTP User Plane, it is talking to a GTP user plane function in the core network which is exactly the same as if you're talking to an S gateway.

The protocol stack relative to the axis networks interface to the core for a user plane does not change while the signaling playing does change as you can see in the image below

For notes on the AMF replaces the MME role towards the Radio Network. S1AP is replaced with the next generation access application part for NGAP.

So, you can take this protocol stack at the bottom and swap NGAP with S1AP (which is what it used to be) which are paired to be almost identical and are both ASN1 encoded binary signaling protocols and doesn't have lots of difference.

The common terminology doesn't change very much as ECGI becomes NCGI for the cell IDs tracking areas. QCI concept is replaced with

Cyrus Jackson

VQI mapping. QCI is a QOS profile in LTE which is an important point because it wasn't done very well when UMTS moved to LTE with four classes moving intoQCI1 through 9.

Well, QCI through 9 in the LTE model for the Evolve Packet System which is going to be mapped directly verbatim 1 through 9 to the VQI

VQI is the QOS profile concept for the 5G core which supports a larger number of profiles but QCI1-9 profiles maintain the same properties as QCI1-9 in the QCI model which makes mapping intuitive.

With this 1 is 1, 2 is 2, and 3 is 3 which means you don't have to think about it more than that.

If you're looking at this from a volte scenario, QCI5 is now 5QI5. QCI1 for voice is now 5QI1 for voice.

CHAPTER 15

CHARGING INTERFACES

Charging interfaces and policy control are two areas which to me are the least changed additions in these chains:

They're the least updated. They're the most to be determined in terms of when you get into the nuts and bolts protocol specs.

So a TS 32 to 91 is the spec for charging but t it needed the full level of detail that you have for say TS 32 to 99 for the GUI interface and for RO interface more generically that equivalent in 3GPP at least at the time of writing this book.

The online charging server and the offline charging server of the previous architecture which are both being depreciated and replaced generically with the idea of a charging function.

P gateway as it relates to charging is being replaced with the session management function. The GUI interface is deprecated. Diameter charging in general goes away in 5GS. So,

Cyrus Jackson

chart diameter is gone and that includes diameter charging.

This is an HTTP 2 interface as you should be able to know just by the fact that it's the letter N followed by the lowercase CHF which always means it's an HTTP 2 interface.

PCC requires HTTP to stack which enables it to get rid of residue in favor of HTTP2. Policy control function is going to be HTTP2 which is something you would need to keep in mind.

If you're looking at all these Network functions and trying to figure out what you need to do or what you need to add or what you need to change about your existing network functions to support this architecture, PCF is definitely one of those things that would need to support HTTP2.

If you have an existing policy control infrastructure based on diameter, which means HTTP2 would eventually be needed.

CTS is the generic concept of a charging client similar to how in the Evolved Packet Core, the P gateway, alternately named the PCEF when

The 5G Network Architecture

you were talking about a policy control enforcement.

Well, the CTS is that which queries the charging function and that's what it really means.

Some diagrams will show SMF while some will show CTS and some others will show SMF with a little box called CTS inside of it but CTS is just specifically the charging client.

The nuts and bolts are TBD but my impression at least is that it they're mostly just trying to take the attributes and values of GY and graft them into the charging function and SMF exchanges and the same sort of deal with the N7 interface in the N5 for the GX and GY respectively from the policy control function.

So, what you would see on N7 in HTTP2 would very much be the same you would see if you're looking at a GX interface from a P gateway to the PCRF.

S9 was originally defined as a PCRF to PCRF interface but there are known deployment in S9 which means they didn't even bother to work out interworking between S9 and the HB2

Cyrus Jackson

equivalent. In other words, S9 is actually deprecated at this point.

The 5G Network Architecture

CHAPTER 16

ROAMING SERVICES

It should be too aware that the security function is defined by 32BB step and it acts as an entrance point into the PLM in the 5G network basically.

So NRF2 to NRF2 traffic which I'd drawn earlier is just a direct line between the two operator interrupts.

Realistically, you would be going through a security gateway to do that most likely.

Most operators would have would have some diameter routing infrastructure for roaming in LTE where that role is really when they are places where they should be.

It does beg the question of how you do routing of HTTP messages or the concept of HTTP2 or HTTP proxy.

With use of GTP user plane, the user plane basically doesn't change between operators so

Cyrus Jackson

the fact that it's just the GTP packet containing an IP packet between two user plane functions.

Well, the difference between sending that between two user plane functions and a user plane function than a P gateway or S gateway is nothing because they are exactly the same thing. It really hasn't changed since the GPRS core, they also use GTP user plane.

CHAPTER 17

DNS ENCHANEMENTS

If you look at TS29, 303 which is the DNS spec for gateway selection in the Evolve Packet Core, they are actually adding user plain functions surface information in the DNS records for the Evolved Packet course specifically.

So, that is something you may need to look at if you're looking at deploying 5G core or looking at 5G roaming.

Here's a cheat sheet for some of the equivalent terms especially in the Evolve Packet Core which you need to:

- **5GC means 5G Core Network.**
- **5G-EIR means 5G Equipment Identity Register.**
- **AKA means Authentication and Key Agreement.**
- **AS means Access Stratum.**
- **AMBR means Aggregate Maximum Bit Rate.**

Cyrus Jackson

- ANDSP means Access Network Discover and Selection Policy.
- AF means Application Function.
- AMF means Access and Mobility Management Function.
- AUSF means Authentication Server Function.
- CS means Circuit Switched.
- CBCF means Cell Broadcast Control Function.
- CHF means Charging Function.
- DC means Dual Connection.
- DNN means Data Network Name.
- EPC means Evolved Packet Core.
- EAP means Extensible Authentication Protocol.
- eMBB means Enhanced Mobile Broadband.
- E-UTRAN means Evolved UTRAN.
- GTP means GPRS Tunneling Protocol.
- HSS means Home Subscriber Server.
- IMSI means International Mobile Subscriber Identity.
- IWF means Interworking Function.

The 5G Network Architecture

- IMSI means International Mobile Subscriber Identity.
- I-NEF means Intermediate NEF.
- I-SMF means Intermediate SMF.
- MAP means Mobile Application Part.
- MioT means Massive IoT.
- MME means Mobility Management Entity.
- MTC means Machine Type Communication.
- NRF means Network Repository Function.
- NSIID means Network Slice Instance Identifier.
- NSSAI means Network Slice Selection Assistance Information.
- NWDAF means Network Data Analytics Function.
- NSSF means Network Slice Selection Function.
- NAS means Non-Access Stratum.
- NEF means Network Exposure Function.
- NF means Network Function.
- NG means Next Generation.
- NB means Node B.
- NR means New Radio.
- OCS means Online Charging System.

- QoS means Quality Of Service.
- PSA means PDU Session Anchor.
- PCF means Policy Control Function.
- PCRF means Policy And Charging Rules Function.
- PGW means Packet Data Network Gateway.
- RAN means Radio Access Network.
- RAT means Radio Access Technology.
- RRC means Radio Resource Control.
- SUCI means Subscription Concealed Identifier.
- SUPI means Subscription Permanent Identifier.
- SEPP means Security Edge Protection Proxy.
- SD means Slice Differentiator.
- SMF means Session Management Function.
- S1AP means S1 Application Protocol.
- SCEF means Service Capability Exposure Function.
- SGW means Serving Gateway.
- SMS-C means Short Message Service Centre.

- SRVCC means Single Radio Voice Call Continuity.
- UCMF means UE Radio Capability Management Function.
- UDSF means Unstructured Data Storage Function.
- UE means User Equipment.
- URLLC means Ultra-Reliable and Low-latency Communication.
- UDM means United Data Management.
- UDR means United Data Repository.
- UDSF means Unstructured Data Storage Function.
- UPF means User Plane Function.

CHAPTER 18

IMS CONSIDERATION

The 5G system is abstract from IMS and they haven't really touched it.

If you're thinking of what comes into the 5G IMS Network, the answer is IMS hasn't really changed because there's still mobility management procedures that are relevant to it like the VOPS flag and the emergency flag in the Evolved Packet Core, you get a consider voice and emergency support on the access side and the AMF side just as I showed you in the MME which play in the IMS a bit.

The 5G system QOS types, as I said in earlier chapters, map intuitively to what work you see in high values for volte which means 5 is 5 and 1 is 1 if you're following the iron 92 recommendations which I assume everyone reading this book does.

CHAPTER 19

DEPRECIATION OF DIAMETER IN 5GS

It is possible ECB could come in and replace Diameter in the nearest future for the IMS Network which is at least 16 plus of amendment.

It's going to be there at least 16 and beyond and may look like a deprecated diameter or at least give an option of HTTP2 in the IMS Network instead of diameter because IMS is the only thing left that's supporting diameter at the end of this at the end of release 15 stage 3.

SMS over SIP is basically unchanged if you're worried about this from an SMS perspective.

A SIP-based SMS is still supported but the only thing to really keep in mind is that there is going to be no such thing as a combined attached in the 5G core on the mobility side and what that's going to mean is the content of attaching to a VLR for circuit-switched SMS is really a thing that's essentially replaced with the ability of an AMF to invoke an SMS function or SMSF to relay NASS.

Cyrus Jackson

So, NASS base SMS is supported in 5G core because you can do it towards an AMF just as you could towards an MME and you can relay it to an SMS Center except that you have this new kind of function called the SMSF which would really be acting as an HTTP2 to SMS PDU converter to get that into the circuits switch domain.

In other words, that could just end up being part of a feature that's supported by an SMS Center and also inherently to support an HTTP2 interface.

Location data changes to NCGI and there are more bits to add a lot larger cell ID than the EPS cell global identity had and that should be reflected in panning it.

So, on a SIP level you have P access networking info as a SIP header and that obviously need to include cell data for the new type of location data as opposed to the old which is something to keep in mind from a charging perspective and from an IMS service support perspective.

However, a lot of things care about location and there is new location data.

Cyrus Jackson

CHAPTER 20

DROP AN HONEST REVIEW FOR THIS BOOK

I'm curious and I'll love a feedback from you:

- How did this book deliver on its promises?

- And what kind of doubts did you have before starting to read this book?

Simply let me know by leaving an honest review on Amazon. I love getting feedback from readers in order to serve you better.

Even if it's a few lines, I'll really appreciate it!

Reviews are very important on Amazon as they speak volume of how the book delivered, shows social proof and give others an idea of the content of the book.

So, if you benefited from this book: **drop a review today** by following this link.

Thank you, I wish you success in all your endeavors.

Cyrus Jackson

Printed in Great Britain
by Amazon